Research Methods and Survey Applications

Outlines and Activities
from a Christian Perspective

Also by David R. Dunaetz

Personality and Conflict Style: Effects on Membership Duration in Voluntary Associations. Lambert Academic Press.

The Early Religious History of France: An Introduction for Church Planters and Missionaries. Martel Press.

Research Methods and Survey Applications

Outlines and Activities from a Christian Perspective

David R. Dunaetz

Martel
Press

Claremont, California

First Published in the United States of America by
Martel Press
Claremont CA

Correspondence:
695 E. Bougainvillea St.
Azusa, CA 91702

editor@martelpress.com

21 20 19 18 17   7 6 5 4 3 2

ISBN: 978-0-9986175-0-3 (print)
ISBN: 978-0-9986175-1-0 (electronic)

Library of Congress Control Number: 2017901916

The images of Tux the Penguin are public domain: Creative Commons 0
(creativecommons.org/publicdomain/zero/1.0/deed.en)
Retrieved from pixabay.com

Scripture quotations taken from the New American Standard Bible®,
Copyright © 1960, 1962, 1963, 1968, 1971, 1972, 1973,
1975, 1977, 1995 by The Lockman Foundation
Used by permission. (www.Lockman.org)

Contents

Preface

This book can be used as a supplement to any traditional Research Methods textbook in the social sciences that focuses on quantitative methods. It is essentially a set of lecture notes, discussion questions, and activities covering the foundational material in a Research Methods course. Each page is designed to help students follow a 20-60 minute segment of class that may include a lecture, a large group discussion, a small group discussion, and/or an activity. The activities are especially important because they provide an immediate opportunity to put into practice key concepts and increase one's research skills.

The target audience consists of students at Christian universities that value integrating biblical concepts into the academic material. Research Methods is not a course that easily lends itself to faith integration because the criteria for good research from a Christian perspective are very, very similar to the criteria for good research from a non-Christian perspective. However, there are occasionally issues where a biblical perspective can shed light on research methods, and vice-versa. Several of the outlines and discussion questions address these issues.

For most students, Research Methods is not the most anticipated course in their degree program. To make the course more interesting and less threatening, the classroom can be organized so that students sit in groups of 3-6 and remain together in these groups throughout the semester. This book assumes that most of the discussions take place in these small groups and that each group is responsible for carrying out a research project during the semester.

Research Methods can also be organized as a service learning class. If students work with a real-life organization to respond to a research question that is pertinent to a problem faced by a decision maker in the organization, the relevance of the research process increases dramatically. For this reason, there is a large emphasis on survey research in this book; survey research is the most common way that research is carried out in organizations, for both ethical and practical reasons.

Chapter 1: Research Basics

Google Scholar

Setting Up Your Computer to Download Articles from the Library

Google Scholar is a great way to find scholarly journal articles and books. Sometimes the full text of an article or book will be available directly from a web site, but often it will not be and you will have to go through your school library to download it. Before you begin searching, it's a good idea to set up your preferences so that you will be able to tell if your school's library can provide access to the article or book you are looking for in Google Scholar. Here's how to set up Google Scholar:

Step 1. Go to the Google Scholar homepage at http://scholar.google.com.

Step 2. Click on ⚙ Settings in the upper right corner.

Step 3. Click on the tab "Library Links" on the left side of the screen.

Step 4. Search for your school (e.g., Azusa Pacific).
Then check both boxes ("Open WorldCat" and "Azusa Pacific University – APU Library FT Finder") and Save.

Now you are ready to start searching, and you will see a link every time there is an article or book which is available either through the APU online databases or in print version at one of the APU libraries. (You may also see PDF versions of article available online from other sources that do not require library access.)

The number "Cited by" can help you quickly determine a rough measure of an article's importance.

Google Scholar

perceived organizational support

About 2,330,000 results (0.03 sec)

Articles

Case law

My library

Reciprocation of **perceived organizational support**.
R Eisenberger, S Armeli, B Rexwinkel... - Journal of applied ..., 2001 - psycnet.apa.org
Abstract 1. Four hundred thirteen postal employees were surveyed to investigate reciprocation's role in the relationships of **perceived organizational support** (POS) with employees' affective **organizational** commitment and job performance. The authors found
Cited by 9456   Related articles   All 26 versions   Web of Science: 380   Import into EndNote   Save   More

[PDF] researchgate.net
APU Library FT Finder

This article is cited by over 9000 other sources. That's a lot. It's probably pretty important.

A pdf of this article is available free at researchgate.net

This article is available at the school's library. A login is required once a day.

When you reach the article, **always download the pdf**. Do not just read the web page. The web page does not include the tables and figures, and it is often poorly formatted.

Common Writing Problems

Abbreviations Used in This Course

A   Awkward (The sentence structure or vocabulary doesn't communicate effectively.)

APA   APA formatting problem

C   Consistency needed (Plural and singular pronouns cannot be used interchangeably; verb tenses must match up; grammatical structures should be parallel)

F   Flow is not clear (The link between adjacent ideas is not clear).

Fr   Fragment (The sentence is not complete; the subject, verb, or a main clause is missing.)

G   Grammar problem

L   Logic is not clear.

RO   Run on sentence (The sentence is too long; break it into smaller ones.)

V   Vague (Be more specific so the reader knows exactly what you are saying.)

W   Wordy (Shorten and remove extra words.)

WC   Word choice seems incorrect.

Common APA Issues

If a number is the first word in a sentence, it must be written out in words (e.g., "Seventy-seven participants").

Abrv   The first time an abbreviation is introduced, it must be immediately preceded by the word or expression spelled out fully (SOF). Once you introduce an abbreviation, you must use it again in the paper and you can no long use the word or expression SOF.

Xline Word, by default, often puts an extra line between paragraphs or at the bottom of the page to avoid widows and orphans. You must turn these features off for APA formatting.

      1. Go to Home => Paragraph => Indents and Spacing => Spacing After to turn off the extra line after paragraphs.

      2. Go to Home => Line and Page Breaks => Pagination => Widow/Orphan Control to turn off widows and orphans.

Latin   Only use Latin abbreviations in parentheses (e.g., "i.e." etc.). The exception is "et al." ("and others") which can be used for a list of authors the second and following time that they appear. It should also be used whenever a work has 6 or more authors.

The 10 Commandments of APA Formatting

1. You need a cover page which has the paper name, your name, and the school name on separate lines, centered and in the middle of the page. To center a line, press Ctrl-E.

2. The entire paper (Header, Title Page, Abstract, Body, References) needs to be in 12 point Times New Roman font with 1 inch margins. Do not use any color other than black.

3. Everything needs to be double-spaced. At the beginning of the body of your text, go to "line spacing" (the button with horizontal lines and two arrows) and select 2.0. Do not use "Enter" except at the end of a paragraph.

4. You need to have a header with the title of the paper in all caps (perhaps shortened if necessary) and the page number. Search for "headers" in "help" in Word. This will make the header repeated automatically on each page with the correct page number; however, you need to make the header on the first page different from the header on the other pages (check the appropriate box) No lines, decorations, abbreviations, or colors are allowed in the header. The header should be in 12 pt Times New Roman. The header on the cover page should begin with the words "Running head:"

5. The body of the paper begins on page 3 (or 2 if there is no abstract). The first line of text (below the header) should be the title of the paper centered.

6. No blank lines are allowed in the body. Word may be set to add an extra line after each paragraph; if it is, you need to turn it off. Go to "line spacing" (the button with a bunch of horizontal lines and two arrows) and choose "Remove space after paragraph" for the whole paper.

7. Paragraphs must be indented ½ inch. Do this with "Tab" at the beginning of each paragraph.

8. In Microsoft Word, whenever a word is underlined with a red, green, or blue squiggly line, it means that there is probably a problem with spelling or grammar. You need to right click on the word to see what Word suggests. You need to fix all the problems before submitting a document.

9. You need to have a references page. The first line on the References page should be "References" centered. Use Ctrl-T at the beginning of a reference to get each one to indent properly (reverse indenting).

10. See the document "Examples and Rules for APA References" (on the following pages) to determine the correct format for your inline references and the references on your References Page.

Rules for APA References

1. The title of the References Page should be "References," centered at the top of the page.

2. Everything should be double-spaced in 12 pt, black, Times New Roman.

3. There should be no additional blank lines. Word may be set to add a space after each paragraph; if it is, you need to turn it off. Go to "line spacing" (the button with a bunch of horizontal lines and two arrows) and choose "Remove space after paragraph" for the whole paper. I suggest adjusting your default settings in Word.

3b. To get the correct indenting, begin each reference with Ctrl-T.

4. The basic format for each reference on the reference page is:

Author (Year). Title. Source, Page Numbers.

See rule 11 to determine when the page numbers are necessary.

5. The author should be listed last name first followed by his or her initials:
Example:

Smith, D. R. (2009). *Endless rules for clear communication*. Los Angeles, CA: Sage.

6. Only the first word of the title of a book or article title is capitalized (but see rule 19 for exceptions). Example:

Dahl, R. (1963). *Charlie and the chocolate factory*. New York, NY: MacMillan.

7. In general, the source (see Rule 4) is either
A) the publisher of a book (preceded by the city and state in which it is published),
B) the name of a journal or a magazine (plus its volume number and pages where the article is found), or
C) a web site.

8. In the References Page, the references need to be alphabetically listed by the last name of the author.

9. Avoid using the References tools in Microsoft Word. It does not follow APA rules very well.

10. If no author is listed, use the name of the organization that produced the document or web page as the author. In web documents, if the name of the organization is not given in the web site, use the main address of the website (for example, everest.com) or simply begin the reference with the name of the article. If no date is available, put n.d. for "no date."

Example:

> *Making the most out of ambiguity*. (n.d.). Retrieved from
>   http:/www.funnyarticles.com/Making_the_most.html

For the inline reference, you can use the abbreviated title as the author. For example: ("Making the most", n.d.)

11. If there are no page numbers, or if the entire work is written by one author, do not include page numbers.

12. For the in-text references, in general, put the author's last name and date of publication in parentheses, for example: (Smith, 2008). See the page *Sample In-Text References* for cases where the author is mentioned in the text or there is a direct quotation.

13. The abbreviations for states are always 2 capital letters and must be used. Publishers must be identified as: City, State: Publisher's name.

14. When a magazine article or any other kind of article is retrieved from the web and no volume number is available, the URL needs to be included in the reference (see 16).

15. When no date is available for printed material or a web site, use (n.d.) for "no date."

16. When citing a web source, precede the address with "Retrieved from". Provide the whole URL, not just the home page (e.g., "Retrieved from http://www.CareerStrategy.org/assessment").

17. Always end a reference with a period, except when it ends with a URL (web address).

18. For the dates in parentheses, include only the year (except for periodicals/ magazines with no volume number; you should include the month or season that it was published).

19. Titles of books and articles only have capital letters for the first word of a title, for the first word after a colon (:) or other punctuation mark, or for proper nouns. Titles of books are written in *italics*. Example:

> Carson, D. E. (1976). *Trip to Zambia: A wild adventure*. New York, NY: Wild Books.

20. Article titles from magazines and journals should be in regular typeface (not italics). Article titles from web pages should be in italics. Journal titles and magazine titles should in italics, with the first letters of the important words capitalized, followed by the volume number in italics if available.

21. If there is a direct quotation for your inline citation, include the page number when the reference is cited, for example: (Kerr, 2008, p. 234). When there is not a direct quotation, do not include the page number.

22. Remove hyperlinks by right clicking on colored text and following the directions.

Sample In-Text References

<u>Sample Paragraph with In-Text References</u>

Psychosocial social treatments may be very effective (Abel, 2000). According to the U.S. Bureau of Labor Statistics (2007), the need for actors is not likely to grow. Doherty (2006, p. 27) claims, "We must be more responsible." He later adds, "We can't keep making the same mistakes over and over" (p. 36).

―――――――――――――――――――

List of Sample in-text citations (typical form):

(Abel, 2000)

(Bureau of Labor Statistics, 2007)

(Doherty, 1996)

(Erisman, n.d.)

(Frankena, 1973)

An Example of an APA Reference Page

References

Abel, E. M. (2000). Psychosocial treatments for battered women: A review of empirical research. *Research on Social Work Practice, 10*, 55-77.

Adams, J. S. (1965). Inequity in Social Exchange. In L. Berkowitz (Ed.), *Advances in Experimental Social Psychology* (Vol. 2, pp. 267-299). New York, NY: Academic Press.

Bureau of Labor Statistics (2007). *Occupational outlook handbook, 2008: Actors, producers, and directors.* Retrieved from http://data.bls.gov/cgi-bin/print.pl/oco/ocos093.htm

Doherty, W. J. (1996). Soul searching: Why psychotherapy must promote moral responsibility. New York, NY: BasicBooks.

Erisman, M. (n.d.). Lecture notes for PPSY 531 moral identity formation and psychotherapy. Azusa Pacific University.

Frankena, W. (1973). *Ethics* (2nd ed.). Upper Saddle River, NJ: Prentice Hall.

Magazine Article Guidelines (To Avoid Being Boring)

Writing for a magazine (whether a popular magazine aimed at a general audience or a trade journal aimed at a specific audience) requires a different writing style than formal academic writing (e.g., theses and journal articles). Here are some guidelines to help you communicate more effectively.

1. Start (and end) with an interesting story to **develop human interest**.

2. Illustrate your points with real or imaginary **stories and examples**. You might want to use a common story/theme/character throughout your paper.

3. **Avoid vague, abstract descriptions** that might bore the typical reader.

4. Use at least a few **emotional words** so the reader knows how to interpret what you're describing.

5. Focus on **persuading**, not on simply providing information. Be clear about what you want people to believe or do.

6. **Use section headings** (unless your style really doesn't call for them, e.g., a humorous narrative). Clearly identify at the beginning of the section what question or issue you are addressing. At the end, recap the question and answer and then transition into the next section.

7. **Avoid technical abbreviations and vocabulary** unless you can use them to make your arguments more convincing.

8. **Avoid numbers** unless they add to your argument. Use them sparingly.

9. Be specific when talking about costs and benefits.

10. **Know your audience**. Only write about things that would interest them, in a style that would interest them.

11. Make sure you are following the **style guide provided by your editor**. Please include it with your paper. If it contradicts APA rules, follow what your editor says.

12. When talking about studies, people care more about **the researchers** than they do about the title of the study. Do not list researchers' names without providing some background information about them. Look up their biographies and make them seem like interesting people.

13. **Organize your paper around questions** you are answering or topics that you cover, not by articles or studies.

14. Be sure to choose a **catchy title**.

15. When you say something **controversial**, be sure to provide **strong arguments** to support your point and respond to **potential objections** that the reader may have.

Chapter 2: The Research Process and Ethics

The Research Process: Applied Research

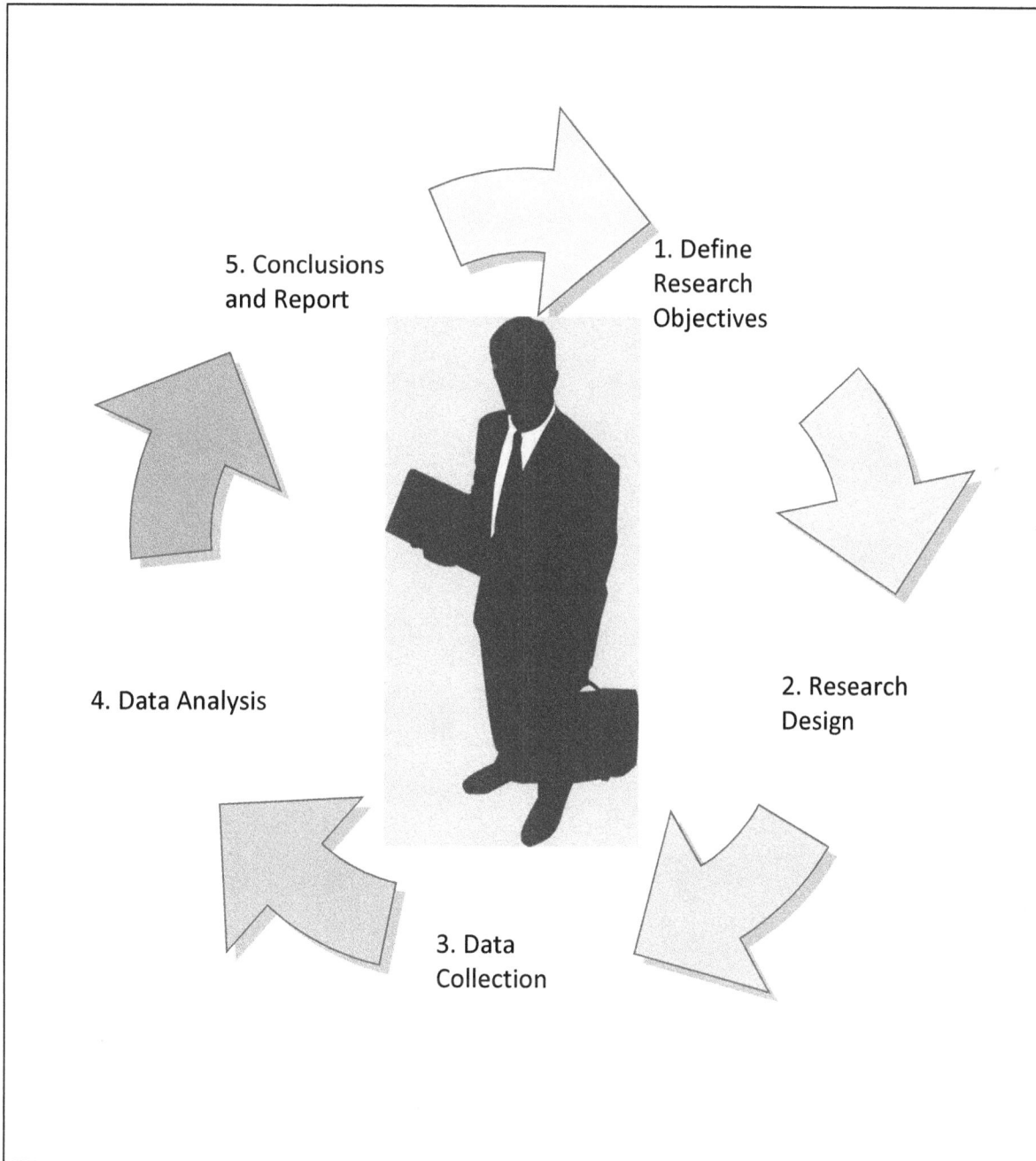

5. Conclusions and Report

1. Define Research Objectives

4. Data Analysis

2. Research Design

3. Data Collection

Flowchart of the Research Process

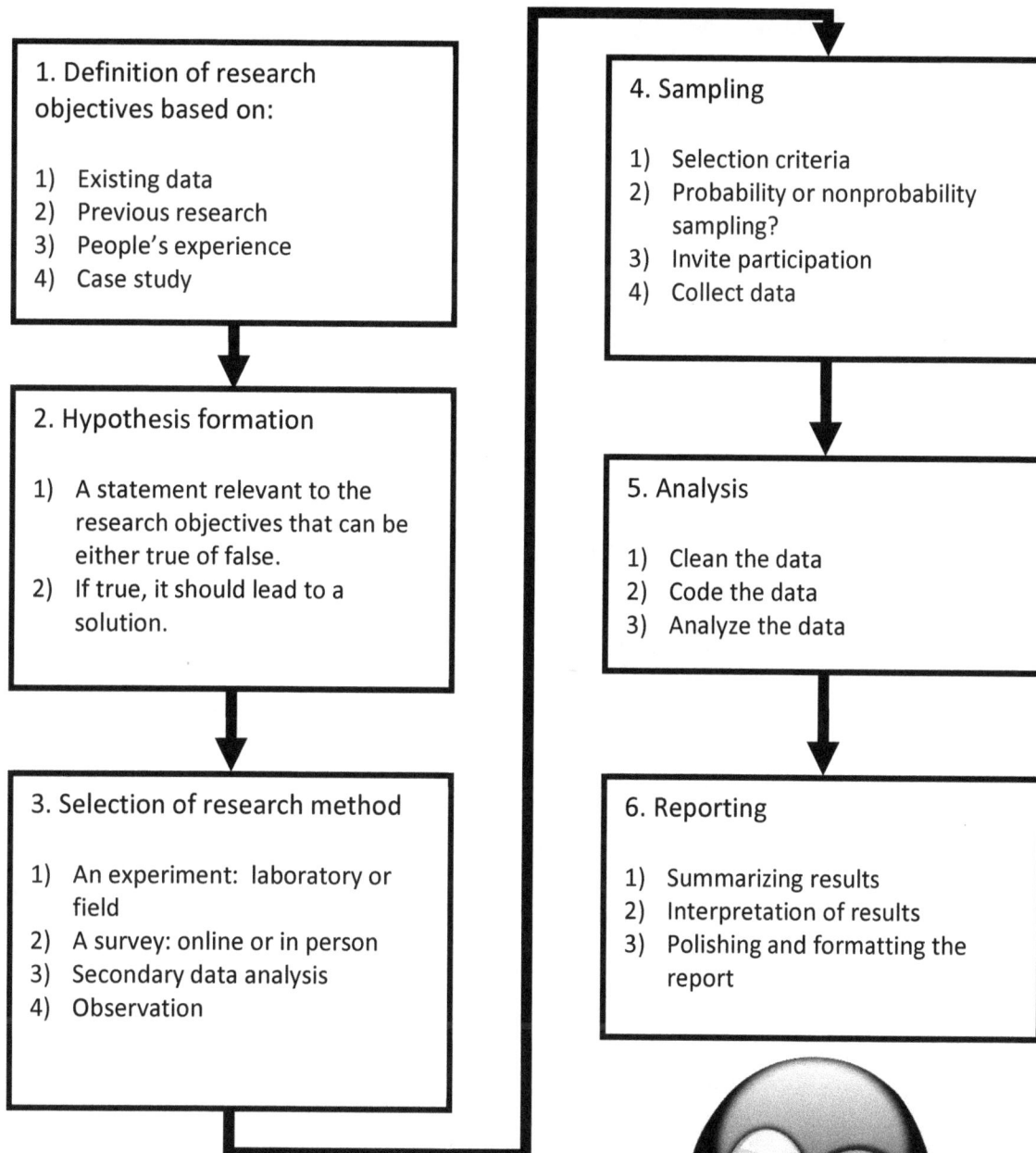

1. Definition of research objectives based on:

1) Existing data
2) Previous research
3) People's experience
4) Case study

2. Hypothesis formation

1) A statement relevant to the research objectives that can be either true of false.
2) If true, it should lead to a solution.

3. Selection of research method

1) An experiment: laboratory or field
2) A survey: online or in person
3) Secondary data analysis
4) Observation

4. Sampling

1) Selection criteria
2) Probability or nonprobability sampling?
3) Invite participation
4) Collect data

5. Analysis

1) Clean the data
2) Code the data
3) Analyze the data

6. Reporting

1) Summarizing results
2) Interpretation of results
3) Polishing and formatting the report

Research Questions and Hypothesis Formation

Research Problem: A problem for which you want to find a solution through research.

Research Question: A question whose response would help solve the research problem.

Hypothesis: A statement that is either true or false. If it is true, then it can help provide a solution to the research problem.

Research Problem	Research Question	Hypothesis
1. People are anxious in stats classes.	Does sitting in groups, rather than rows, reduce anxiety?	Students who sit in groups will be less anxious than students who sit in rows.
2. Students quickly forget the material presented in stats classes and are not able to do their homework.	What methods can be used to increase stats learning?	Students who attend a stats class and watch videos will do better on their homework than students who just attend class.
3. Employees often forget to come to department meetings	Why do employees forget to come to department meetings?	Emailing an agenda the day before a department meeting will increase attendance.
4. Online advertising is not always effective.	Do online ads influence the buying behavior of teenagers?	Teenage boys will buy more of the products advertised online than teenage girls will.

Characteristics of Good Hypotheses
1. A response to a research question stated in a declarative form.
2. Posits a relationship between 2 or more variables.
        -A difference between groups (e.g., height of males and females)
        -A correlation (Age and height of senior citizens)
3. Reflects a theory or body of literature upon which it is based.
4. Brief and to the point.
5. Testable.

Discussion

1. Evaluate the research questions and hypotheses in examples 3 and 4.
2. For your team project, how will you determine what an appropriate research problem is?
3. Imagine a possible research question. What would be an appropriate hypothesis?
4. How might your research problem relate to God's purposes?

The Literature Review

Purpose
1. The literature review responds to the research question or problem by summarizing what is known about the topic.
2. Its logical conclusion should be your hypothesis.

Traditional American Approach
I.   Introduction
    A.   Attention getting general statement
    B.   Description of the problem
    C.   Tentative thesis statement = tentative hypothesis
II. Details of the Solution
    A.   Solution point 1
    B.   Solution point 2
    C.   Solution point 3
III. Conclusion
    A.   Restatement of problem
    B.   Summary of solution = hypothesis

Traditional European
I.   Thesis (an important idea)
II. Anti-thesis (an idea that seems to contradict or nullify the thesis)
III. Synthesis (how the thesis and the anti-thesis can be combined in a coherent whole) = hypothesis

Note 1: The literature review is not a summary of one article after another. It is a clear argument leading up to your hypothesis.

Note 2: Make sure you do not make assumptions with which your audience may not agree. This lowers your credibility. For example, you could cite the Bible freely for an evangelical audience, but for a more general audience, you should not assume the reader will accept the Bible as an authoritative source.

The Literature Review: 3 Types of Sources

1. Start with general sources to get ideas.

2. Look at several secondary sources to get an overview of the topic.

3. Study the primary sources to master your subject.

Type of Source	Purpose	Examples for Organizational Psychology
General Source	Provides an overview of the topic in an easy to read way	Wikipedia Textbooks Newspapers Harvard Business Review
Secondary Source	Provides a scholarly analysis or summary of the relevant research done on the topic	Handbooks Meta-analyses Review articles Books summarizing a researcher's research program
Primary Source	Provides original evidence of the ideas relevant to the topic	Peer-reviewed empirical articles Theoretical articles

4. Always keep very organized notes (e.g., an annotated biography or outline of each article)

5. Write your proposal once you have mastered the topic and know what is missing in the literature.

The Literature Review and Hypothesis Formation

This is how we tend to think of the relationship:

Research Question ➡️ **Hypothesis** ➡️ **Literature Review**

This first approach is too linear; it doesn't work. Your literature review will inform your research question and your hypothesis.

This is a much more realistic view of the relationship:

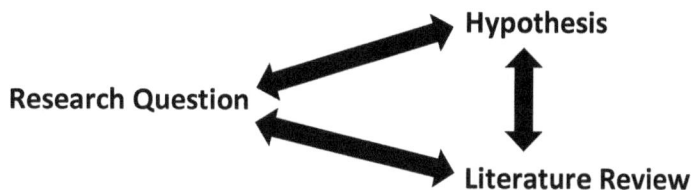

Hypothesis

Research Question

Literature Review

This approach shows how your Research Question and Hypothesis might change as you do your literature review.

When finished, your literature review should read like an essay leading up to your hypothesis as a conclusion.

> A. An important question is "How does A relate to B?"
> B. We know this about A.
> C. We know this about B.
> D. Therefore, we hypothesize that A and B are related in such and such a way.

Literature Review Scavenger Hunt

After choosing a topic, each team will race to collect the following information to form a lit review.
<u>Scoring</u>

-The correctly formatted APA reference for each item is worth one point.

-Download the .pdf of the source for an extra bonus point for each item.

-Send each item (and potentially the corresponding .pdf) in a separate email to your professor and **all team members. Your team name and Item number (1-7) must be in the subject line.**

<u>General Sources</u>

☐1. Find the **Wikipedia** article most relevant to your topic.

☐2. Find a section in **a textbook** that you own which deals with your topic. Use Amazon or Google Books to search through it. Paste a snippet (or screen shot) of the page into your email.

☐3. Find **a book** (at the library or Amazon) providing **an elementary introduction** to your topic on the popular level.

Relevant data bases at APU Library Site: -APU Collection (books at APU)

-Link+ (books at nearby libraries which can be delivered to APU)

<u>Secondary Sources</u>

☐4. Find **a lit review** (or a meta-analysis) journal article relevant to your topic on Google Scholar (search "review + your topic").

☐5. Find **a book** (on Amazon or at the library) written by a researcher that summarizes research relevant to your topic or treats it in an in-depth way (beyond the textbook level).

<u>Primary Sources</u>

☐6. With **Google Scholar**, find a **research article that reports empirical data** and results relevant to your subject.

☐7. With **Google Scholar**, find a **theoretical article** that describes in depth a theory relevant to your topic.

IRBs and Certification for Protecting Human Participants

As Christians, we want to uphold the highest ethical standards in order to serve others and glorify God.

As researchers, we need to follow government regulations and become certified in protecting people when they participate in research.

<u>Key Concepts</u>
National Institutes of Health (NIH)
Informed Consent
Deception
Government Regulations
Vulnerable Populations: unborn, children, mentally disabled, prisoners, etc.
Institutional Review Board (IRB) approval needed for various types of research:
-Excluded classroom research
> -Low risk research done in Research Methods classes with non-vulnerable populations that will not be published or presented publically.
> -Can be approved by professor without any forms

-Exempt research
> Either:
> A) Low risk research done on non-vulnerable populations that will be used in a public presentation or publication OR
> B) Low risk research done with vulnerable populations when the researchers have no interaction with members of a vulnerable population; data must be anonymous; IRB form needed (plus a few others; see IRB Handbook)

-Expedited reviews: Low risk studies with vulnerable populations, or non-anonymous data (allow 1 month for approval)
-Full reviews: Non-low risk studies (allow 1-2 months for approval)

Certification in Human Subjects Research. Government regulations require you to be certified in Human Subjects Research. To earn this certificate, you need to complete a 2 hour online course:

- Go to https://www.citiprogram.org (or whatever link is provided by your university)
- Create an account using the right-hand panel.
- When Creating an Account, affiliate with your university.
- After registering, access the Humans Subject Research (Social & Behavioral Research - Basic) under your university's Courses tab.

This course needs to be completed early in the semester before you begin collecting data.

Informed Consent Example for Electronic surveys

AZUSA PACIFIC
U N I V E R S I T Y

Informed Consent for [Insert title here]

Voluntary Status: You are being invited to participate in a survey research study. Your participation is voluntary which means you can choose whether or not you want to participate. You may withdraw any time without penalty.

Purpose: The study for which you are being asked to participate is designed to...**[insert a brief explanation of the study]**

Possible Risks: It is expected that participation in this study will provide you with no more than minimal risk or discomfort which means that you should not experience it as any more troubling than your normal daily life. While there are no direct benefits to participating, your response will help us to better understand the research topic.

Confidentiality: The investigator involved with the study will not be collecting any personal information for the study. All responses to this survey are anonymous and confidential. Your name or identity will not be linked in any way to the research data. Concerning your rights or treatment as a research subject, you may contact the Research Integrity Officer at Azusa Pacific University (626) 812-3034.

Consent: I understand that my participation in this study is entirely voluntary and that I may refuse to participate or may withdraw from the study at any time without penalty. I have read this entire form and I understand it completely. By clicking NEXT below and completing the online assessments that follow, I am giving my consent to participate in this study.

Common Writing Problems Examples

In attempting to avoid plagiarism, students often try to change a few words in a sentence. In fact, this is still plagiarism even if the reference is cited. You must always create your own sentence structure rather than use another author's. Use references to show the origin of your ideas. Your sentences should be structured in such a way that they are clear and unambiguous. They must also fit into the flow of your argument.

A   Awkward (The sentence structure or vocabulary doesn't communicate effectively.)

There was no past research done for this organization that can correlate to our research.

APA   APA formatting problem

32 participants completed the survey.

C   Consistency needed (Plural and singular pronouns cannot be used interchangeably; verb tenses much match up; grammatical structures should be parallel)

One participant said they did not want to continue.

F   Flow is not clear (The link between adjacent ideas is not clear.)

Participants completed a survey consisting of 47 Likert items. Mean scores were calculated for all constructs. There were several important variables.

Fr   Fragment (The sentence is not complete; the subject, verb, or a main clause is missing.)

And also take the participants' time to go over them.

G   Grammar problem.

Everyone of the participants indicated satisfaction with the evaluation process.

L   Logic is not clear.

All participants completed the survey indicating that it was well designed.

RO   Run on sentence (The sentence is too long; break it into smaller ones.)

The most common response was "reading and writing" is the most difficult study habit for them to master.

V   Vague (Be more specific so the reader knows exactly what you are saying.)

The multiple choice questions resulted in some commonalities.

W   Wordy (Shorten and remove extra words.)

The first step that we made was figuring out which audience we needed to target to conduct our survey research.

WC   Word Choice seems incorrect.

We chose this style of survey to accumulate the most accurate results.

Common Writing Problems Exercise

Try to fix the following sentences, paying close attention to the problem noted. Some of the sentences provided are so unclear that you'll have to guess what was meant. You may add extra ideas if necessary to improve the sentence.

A   Awkward (The sentence structure or vocabulary doesn't communicate effectively.)

In one known study of adult's perceptions of growing up in long-term foster care, it can be seen that the adoption group was emerging significantly better.

APA   APA formatting problem

Negative consequences to adoption weren't found. (Smith, 1999, p. 46)

F   Flow is not clear (The link between adjacent ideas is not clear).

Previous studies have shown that adopted children suffer fewer negative consequences than children placed in long term foster care. The legal status of children changes when they are adopted.

Fr   Fragment (The sentence is not complete; the subject, verb, or main clause is missing.)

Anxiety and uncertainty in children and their caregivers because of the lack of permanency.

G   Grammar problem

Foster care seems to leave children feeling insecure and lacking a full sense of belonging than did those who were adopted.

L   Logic is not clear.

         Previous studies suggested when it was popular to believe that non-problematic children went into foster care, it was also reported that around 1/3 of them were "disturbed."

RO   Run On sentence (The sentence is too long; break it into smaller ones.)

         Long-term foster care is defined as a type of permanent arrangement, except that the term "permanent" doesn't exactly mean forever because the parental responsibility of the child still often lies with the birth parent or the local authorities.

V   Vague (Be more specific so the reader knows what you are saying.)

         It is reaching that level of passing on rights to the adoptive parents.

W   Wordy (Shorten and remove extra words.)

         Children who were adopted grew up with a well-adjusted life.

WC   Word Choice seems incorrect.

         In general, adopted children undergo fewer problems than foster care children.

Chapter 3: Surveys and Measurement

Theories, Constructs, and Operationalization

Theory

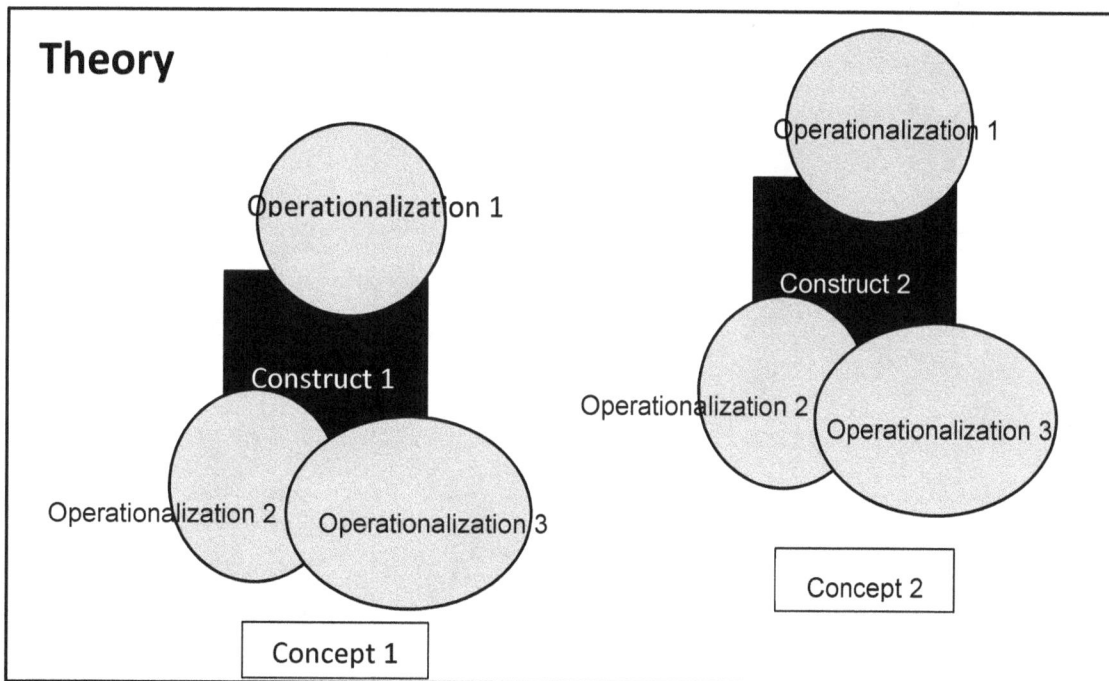

1. Theory: A broad description of how several concepts relate.

Example: **Social Identity Theory**: One's behavior towards a person in another group is influenced by one's own group membership.

2. Concept: A broad idea

Example: **Group membership**: Seeing myself as part of a group.

3. Construct: A variable that describes a specific manifestation of the concept. It varies from one person to another.

Example: **Americanism**: How much do I view myself as an American?

4. Operationalization: A measure of a construct

Example 1: A survey measuring one's belief about being American, often on a Likert scale.
-I feel American.
-I'm more at home in America than any other country.
-I believe in American values.

Example 2: Observations of how often a person wears an American flag pin or flies an American flag.

Example 3: The percentage of time spent in the United States between the ages of 12 and 19.

5 Steps to Questionnaire Design

1) **What should be asked?**
   a) The problem statement and the hypotheses need to be clear.
   b) The relevant variables and constructs need to be clearly understood.
   c) Don't start writing the questionnaire until it is extremely clear what you are looking for.

2) **How should the questions be phrased?**
   a) Questions must be relevant; they should seek only the information needed.
   b) Questions must be clear enough to provide accurate information concerning all the variables and constructs.
   c) If possible, use pre-existing validated questionnaires.

3) **In what sequence should the questions be arranged?**
   a) Earlier questions can influence responses to later questions.

4) **What questionnaire design and layout will best serve the research objectives?**
   a) What type of survey questions will you use?
   b) Will it be administrated in person, on paper, or online?
   c) All ambiguities must be avoided.

5) **How should the questionnaire be pretested?**
   a) The information you gain from the pretest should be used to revise the questionnaire.

Ways of Displaying Questions on Surveys

Multiple Choice/Radio Buttons (Only one choice allowed)

Last month, did you buy something on Amazon.com?

○ Yes

○ No

○ I don't know

Check Boxes (Can choose more than one)

Where have you lived in the U.S.? (Check all that apply.)

☐ The East Coast

☐ The South

☐ The Midwest

☐ The Rocky Mountain States

☐ The West Coast

☐ Overseas

☐ Other:

Linear Scale/Scaling Question

How much do you like Research Methods?

	1	2	3	4	5	
Not at all	○	○	○	○	○	Very much

Dropdown Menu (Avoid unless necessary)

Where do you live?

Choose ▾

Fill in the Blank: Short Answer (Do not make it required).

What did you like most about stats?

Your answer

Fill in the Blank: Paragraph/Long Answer (Do not make it required).

What did you like most about stats?

Your answer

Likert Scale/Multiple Choice in Grid Format

Indicate you agreement with each of the following statments.

	Strongly Disagree	Disagree	Slightly Disagree	Neither Agree nor Disagree	Slightly Agree	Agree	Strongly Agree
I am talkative	○	○	○	○	○	○	○
I am reserved	○	○	○	○	○	○	○
I am full of energy	○	○	○	○	○	○	○
I am quiet	○	○	○	○	○	○	○
I have an assertive personality	○	○	○	○	○	○	○
I am shy and inhibited	○	○	○	○	○	○	○
I am outgoing and social	○	○	○	○	○	○	○

Avoiding Mistakes when Designing Questionnaires

I. **What should be asked?**
   A. Relevant questions
      1. Do they measure the construct or variable we want to measure?
      2. Avoid questions that would be interesting to ask but are irrelevant to your hypotheses.
   B. Questions that provide accurate responses. They should be:
      1. Simple
      2. Understandable
      3. Unbiased
      4. Unambiguous
      5. Nonirritating

II. **What to avoid?**
   A. Complexity
   B. Leading questions (implies a certain answer)
   C. Loaded questions (emotionally charged or socially desirable answer available)
   D. Ambiguity (unclear questions or choices, incomplete range of choices, categories that are not mutually exclusive, etc.)
   E. Double-barreled questions (often contain *and, or, that, which, because,* etc.)
   F. Burdensome questions
   G. Questions that don't generate variance (where most people will mark the same answer)
   H. Inapplicable questions
         -Use skip logic (skip questions) to make sure people only get relevant questions.
   I. Ranking questions with more than 4 choices.
   J. Forced answering with open-ended or controversial questions.
   K. Anything that will lead to missing data ("other," "none of the above," "no opinion")

Survey Question Ordering and Layout

I.   **How to order questions?**
   A.   Start with easy and interesting.
   B.   End with demographics and other personal information.
   C.   Avoid order bias (when unsure, people often choose the 1st choice)
   D.   Ask general questions before specific ones (the funnel technique): "Are you happy with your life?" before "Are you happy with your marriage?"

II. **Layout**
   A.   Start with:
      1.   Good title
      2.   Introduction
         a.   Purpose of study
         b.   What participants will get out of it (appeal to self-interest)
         c.   How long it will take
      3.   Information on anonymity, informed consent.
   B.   Provide clear instructions for each section.
   C.   Use grids (especially for Likert items).
   D.   Likert and fixed alternative questions are quick, easy, and effective.
   E.   Try to use the same scale as much as possible.
   F.   Use lots of white space.
   G.   Use a status bar if useful (often not useful if skip logic is used).
   H.   Proofread and test out skip logic thoroughly.
   I.   Radio buttons and check boxes are good; drop-down boxes are bad.
   J.   Minimize the use of open-ended questions.

Big College Survey

Practice Evaluating Surveys

On the following pages is a survey put together by the Human Resources department of Big College. They're studying what makes a good workplace in order to improve the working conditions at Big College.

Go through the questionnaire and answer the following questions:

     1. Does the survey accomplish its purpose?

     2. What do you think of each of the items?

     3. How could it be improved?

Big College Survey

We're trying to figure out who is happy at work and who isn't. Please fill this out. Itt will only take about 10 or 15 minutes

* Required

1. **What department do you work in?**

   ----

2. **Do you work in administration or are you a faculty member?**
   Mark only one oval.

   ( ) Yes

   ( ) No

   ( ) Both

3. **How long have you worked at your current position?**
   Mark only one oval.

   ( ) 0-1 months

   ( ) 2-12 months

   ( ) Less than 1 year

   ( ) More than 2 years

4. **How did you find out about your current job?**
   Mark only one oval.

   ( ) Online

   ( ) Through a friend

   ( ) Other: _____

5. **Did you know that Big College pays the best salaries of any college in the state?**
   Mark only one oval.

   ( ) Yes

   ( ) No

6. **Please rate what you think is the most important in your job:**
   Mark only one oval per row.

	Very good	Good	Neutral	Bad
Nice office	⬭	⬭	⬭	⬭
Nice people	⬭	⬭	⬭	⬭
Good salary	⬭	⬭	⬭	⬭
Career opportunities	⬭	⬭	⬭	⬭
Benefits	⬭	⬭	⬭	⬭
Catoptric conditions	⬭	⬭	⬭	⬭

7. **Compared to working at other large schools in the state, how satisfied are you?** *
   Mark only one oval per row.

	Very dissatisfied	Dissatisfied	Satisfied	Very satisfied
Big College	⬭	⬭	⬭	⬭
Atlantic School of the Pacific	⬭	⬭	⬭	⬭
Giant University	⬭	⬭	⬭	⬭
Eastwestern University	⬭	⬭	⬭	⬭

8. **If not, what are you dissatisfied about with about your school?**
   Mark only one oval per row.

	Very good	Good	Bad	Very Bad
Big College	⬭	⬭	⬭	⬭
Atlantic School of the Pacific	⬭	⬭	⬭	⬭
Giant University	⬭	⬭	⬭	⬭
Eastwestern University	⬭	⬭	⬭	⬭

9. **Which college pays the best salaries?**
   Mark only one oval.

   ⬭ Big College

   ⬭ Atlantic School of the Pacific

   ⬭ Giant University

   ⬭ Eastwestern University

10. **Can you list the large universities in this state?**

11. **How often do you have contact with people from the following universities?**

   Mark only one oval per row.

	0-1 times per month	2-3 times per month	4-5 times per month	6 or more times per month	0-1 times per year	2-3 times per year	4-5 times per year	6 or more times per year
Big College	○	○	○	○	○	○	○	○
Atlantic School of the Pacific	○	○	○	○	○	○	○	○
Giant University	○	○	○	○	○	○	○	○
Eastwestern University	○	○	○	○	○	○	○	○

12. **Have you ever applied for a job in another university?**

   Mark only one oval.

   ○ Yes

   ○ No

   ○ Which one?

   ○ Other: _____

13. **How do you pay for your expenses?**

   Mark only one oval.

   ○ Check

   ○ Check card

   ○ Credit card

   ○ Revolving charge

   ○ Debit card

   ○ Cash

Statistical questions

The following questions will be kept confidential

14. **What is your age?**

   _____

15. **What is your sex?**

   Mark only one oval.

   ○ Male

   ○ Female

   ○ Prefer not to say

   ○ Other:

16. **What is your annual income?**

_____

17. **What is your marital status?**

Mark only one oval.

○ Single

○ Married

○ Divorced

○ Remarried

○ Widowed

○ Other: _____

Chapter 4. Sampling and Statistics Review

Types of Variables in Experiments and Causal Models

Type of Variable	Meaning
Independent	• A variable that influences the outcome. Potentially a cause of the effect being studied. • It can be changed, adjusted, or manipulated by the experimenter. • Also called a "treatment," "predictor," or "factor."
Dependent	• The effect of a treatment that can be measured. • The result of changes in the independent variable. • Also called "outcome" or "criteria."
Control	• A variable that may or may not influence the dependent variable, but whose influence is removed in the calculations.
Extraneous	• A variable that influences the dependent variable but which is not measured. • Its influence remains unknown.
Moderator	• A variable that modifies the strength of the relationship between the IV and the DV. • It makes the treatment more effective in one condition than in another. • Also known as an interacting variable.
Mediator	• A variable that explains why there's a relationship between the IV and DV. • The IV leads to leads mediator which leads to the DV

Discussion

1. What are the dependent and independent variables in your study?

2. What control variables should you measure?

3. What are some possible extraneous variables?

4. Can you think of any possible moderating variables?

5. What mediating variable might possibly explain a relationship in your study?

Interpreting Correlations

r, the Pearson correlation coefficient, always between -1 and 1.

The Sign of *r*

+   As one variable goes up, so does the other (ex.: height and shoe size)
-       As one variable goes up, the other goes down (ex.: average hours slept and fatigue)

Significance Testing of *r*

If *r* is significant ($p < .05$), then we can be reasonably sure that there is a positive ($r > 0$) or negative ($r < 0$) relationship.

The Strength of *r*

If the absolute value is:	The relationship is:
.8 - 1.0	Very strong (might be measuring the same thing twice)
.6 - .8	Strong
.4 - .6	Moderate
.2 - .4	Weak
.0 - .2	Very Weak

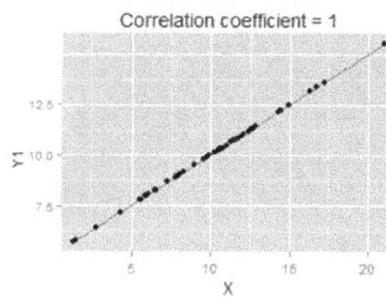

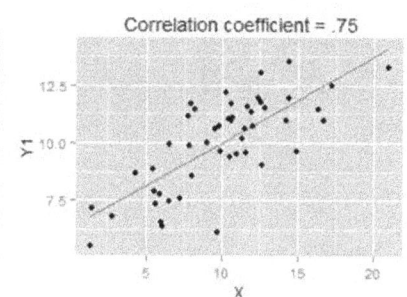

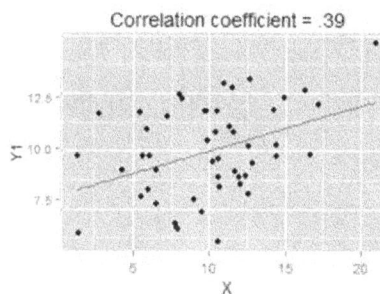

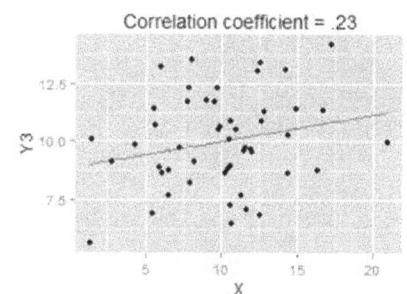

Source: Unknown.

Correlation and Causation

4 Possible interpretations of correlation:

1. A causes B
2. B causes A
3. A third factor causes both A and B
4. The correlation is spurious; it might not exist in a different sample or in the whole population.

Variance (Variation) Explained

r^2 = variance explained in one variable by the other variable (Coefficient of Determination)
$1 - r^2$ = variance explained by other factors (Coefficient of Alienation)

t-Tests and Confidence Intervals

t-tests
A t-test is used to test if the averages of two groups are significantly different.
The higher *t* is, the less likely the difference is due to chance.

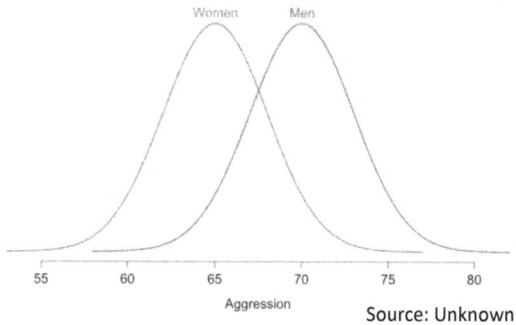

Source: Unknown

Effect sizes of *t*-tests = *d*
d = the number of standard deviations the two means differ.

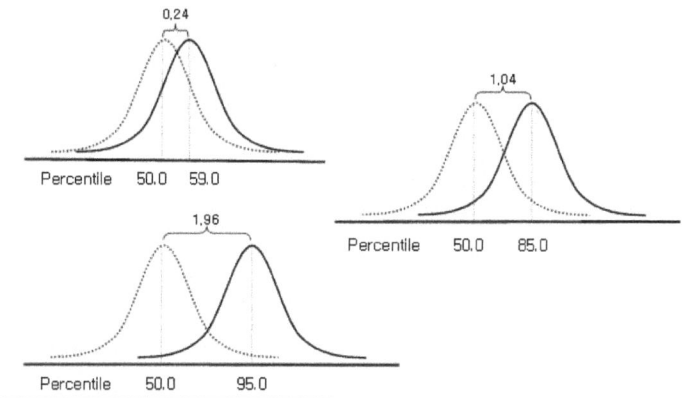

Source: Unknown

Confidence Intervals

A *confidence interval* (*CI*) is a range of numbers that probably includes the mean of the population.

The 95% *CI* means that we can be 95% sure that the true mean of the sample is in that range.

A confidence interval can be presented as [2, 4] or 3 ± 1.
    Note: 3 is the mean, and 1 indicates the range of uncertainty around the mean.
            In Excel, the range of uncertainty is called the "confidence level."

Multiple Independent Variables

The principle of multiple causation states that there can be more than one cause of a given effect.

This means that each DV can have several IVs.

<u>If the IVs are continuous variables</u>, we would use ***multiple regression*** to see which variables significantly predict the DV.

       Example: Which of the Big 5 personality traits predict organizational commitment?

       Each IV will have a β (beta) or B value. It is significant if the probability of getting that value by chance alone is less than 5% ($p < .05$).

<u>If the IVs are categorical (nominal) variables</u>, we can use a ***factorial design*** to test their influence.

       Example: Do gender, education, and age predict supervisor ratings?

           Factors: 1. Gender: Male or Female
                            2. Education: Less than rater, same as rater, or more than rater
                            3. Age: Younger than rater, same age as rater, older than rater

Supervisor Ratings

| | | Age of Supervisor: | | | | | | | | |
| | | Younger than rater | | | Same age as rater | | | Older than rater | | |
	Education of Supervisor:	Less than rater	Same as rater	More than rater	Less than rater	Same as rater	More than rater	Less than rater	Same as rater	More than rater
Gender of supervisor:	Female									
	Male									

Types of Sampling Strategies: Probability and Non-Probability

		Example	Advantages	Disadvantages
Probability Strategies	**Simple Random Sampling**	Putting the name of everyone in the "sampling frame" (e.g., all the employees in an organization) in a hat and randomly selecting the number of participants you need.	Increases internal and external validity: It's a good design and it has good generalizability.	Difficult
	Systematic Sampling	Choosing every 10th person on a list	Pretty much assures random selection, except in rare cases.	Difficult
	Stratified Random Sampling	Choosing every 10th person from a list of male employees and every 10th person from a list of female employees	Makes sure there's enough of each category that's critical to the design of the study.	Difficult
	Cluster Sampling	Randomly selecting a class for each condition of an experiment	Easy	Members of the two clusters may be very different from each other, introducing an extraneous variable.
Non-Probability Strategies	**Convenience Sampling**	Getting as many of your Facebook friends who fit the selection criteria as possible to complete a survey	Easy	External validity (generalizability) may be limited.
	Quota Sampling	Convenience sampling for one or more groups, e.g., 50 males and 50 females	Easy	Same as above

Discussion

1. What is the population you want to study?

2. What are your sampling units? Individuals, dyads, teams, organizations, . . . ?

3. Will you use probability or non-probability sampling? What sampling procedure will you use?

4. What is your sampling frame (the list of people from whom you can draw your sample)?

5. What sample size are you aiming for?
    Use an online sample size calculator such as
        http://www.sample-size.net/correlation-sample-size/

    Based on previous research, what effect size do you expect to find?

Effect Size	d	r
Small	.20	.10
Medium	.50	.30
Large	.80	.50

    How much power ($1 - \beta$, 0-100%) do you want to detect this effect?

Chapter 5. Reliability and Validity

Validity of Research Designs

Man prefers to believe what he prefers to be true.
    -Francis Bacon

There is enough light for those who want to believe and enough shadows to blind those who don't.
    -Blaise Pascal

External Validity

    -The degree to which the results of the study apply to the population that interests us (generalizability)

Threats to external validity
    1. The sample is not representative of the population.
    2. The setting (place, materials, etc.) is not realistic, influencing people's responses.
    3. The historical (time) or cultural context is different in the study and in the application

Internal Validity

    -The degree to which the study was done correctly to test the hypotheses.

Threats to Internal Validity
    1. Confounding variables: other IVs that influence the DVs
    2. Selection bias: choosing a certain type of people who do not accurately represent the population.
    3. History effects: Something in the environment (such as a natural disaster or a major news story) influences people during the testing.
    4. Maturation effects: People mature with time and experience, perhaps changing the results when multiple measures are taken.
    5. Testing effects: Repeatedly asking the same questions or giving the same test may cause people to change their answers
    6. Instrument change: Using more than one version of a survey.
    7. Differential attrition: Some types of people drop out of the study more than others.
    8. Experimenter bias: We tend to see what we want to see, and even influence others to see what we want to see.
    9. Hawthorne effect: We perform better when people pay attention to us.

Discussion

1. What threats to external validity does your study face? What can you do about it?

2. What threats to internal validity does your study face? What can you do about it?

Reliability and Types of Error: Trait and Method

Reliability
-When the test/questionnaire/rater measures the same thing more than once and gets the same outcomes.
-Never perfect.
-Like a correlation: 0 = no reliability, 1 = perfect reliability
-Types: See separate sheet.

Error
= The difference between the true and measured score
⇨   The greater the error, the lower the reliability.

True Score + Error = Observed Score

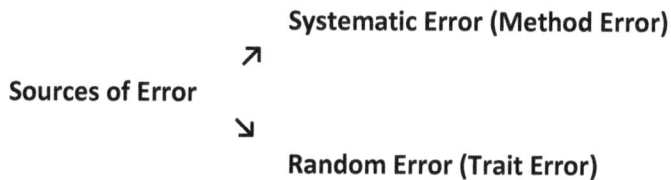

Systematic Error (Method Error)
↗

Sources of Error

↘

Random Error (Trait Error)

A. Method Error:
-Results from how the variable was measured.
-Examples: -The one and only rater is in a bad mood today.
                -People are given different versions of a survey.
                -The control group is measured on an exceptionally hot day; the treatment group on a cold day.
                -A sample is not representative of the population
-Solutions: -Use multiple measures of your construct.
                -Design research so that factors that influence the DV (apart from the IV) are minimized.

B. Trait Error:
-Results from factors that vary within participants.
-Examples: -Some participants are in a bad mood today.
                -Some participants can't understand the survey.
                -Some participants are in a hurry and mark all center scores.
                -How participants understand questions varies from person to person.
-Solution: -Random assignment for experiments and correct design.

Ways to Increase Reliability on Surveys
1. Increase N: Trait error (random error) will cancel out.
2. Make sure every item is clear; measure each variable with several similar items.
3. Make sure all participants respond under the same conditions.
4. Design scales so that the average item is near the midpoint.
5. Make sure everyone receives the same instructions.

Discussion
1. What is your team doing to maximize reliability?
2. What kinds of error do you think risk causing the most problems?

Four Types of Reliability

Reliability is a form of consistency

Type of Reliability	Description	How It Is Measured
Test-Retest	Measures the stability of scores between two points of time within the same participant.	The correlation between responses of Time 1 and Time 2.
Internal Consistency	Measures the degree to which the items measure a single construct	Cronbach's alpha (see below)
Parallel Forms	Measures the degree to which changing the form of the questionnaire changes the responses.	Correlation of scores using Form 1 and Form 2.
Inter-Rater	Measures the agreement between two raters or observers who make measurements	The percentage agreement between the two raters, or the correlation of their scores.

Internal Consistency (Cronbach's alpha) is very important when measuring psychological constructs since most variables can't be measured directly.

    -See the example on the following page of how we can measure extraversion with multiple items.

    -Because extraversion is so broad of a concept, many questions must be asked to reduce trait (random) error

    -Note that some questions are reverse scored.

    -Alpha > .6 is extremely desirable, > .7 is more credible.

Online calculator for Cronbach's alpha: http://www.wessa.net/rwasp_cronbach.wasp

Measuring Extraversion (Big Five Inventory)

I see myself as someone who...

1. ...Is talkative

Strongly Disagree   1   2   3   4   5   Strongly Agree

6. ...Is reserved

Strongly Disagree   1   2   3   4   5   Strongly Agree

11. ...Is full of energy

Strongly Disagree   1   2   3   4   5   Strongly Agree

16. ...Generates a lot of enthusiasm

Strongly Disagree   1   2   3   4   5   Strongly Agree

21. ...Tends to be quiet

Strongly Disagree   1   2   3   4   5   Strongly Agree

26. ...Has an assertive personality

Strongly Disagree   1   2   3   4   5   Strongly Agree

31. ...Is sometimes shy, inhibited

Strongly Disagree   1   2   3   4   5   Strongly Agree

36. ...Is outgoing, sociable

Strongly Disagree   1   2   3   4   5   Strongly Agree

References

John, O. P., Naumann, L. P., & Soto, C. J. (2008). Paradigm Shift to the Integrative Big-Five Trait Taxonomy: History, Measurement, and Conceptual Issues. In O. P. John, R. W. Robins, & L. A. Pervin (Eds.), *Handbook of personality: Theory and research* (pp. 114-158). New York, NY: Guilford Press.

John, O. P., Donahue, E. M., & Kentle, R. L. (1991). *The Big Five Inventory--Versions 4a and 54*. Berkeley, CA: University of California,Berkeley, Institute of Personality and Social Research.

https://www.ocf.berkeley.edu/~johnlab/measures.htm

Validity of Tests and Measures

A test score, instrument, or measure is **valid** if it actually measures what it is supposed to measure.

   -The root meaning of *valid* is strong. Validity describes how strong your research method or your measure is.

   -Validity goes from low to high, but can't be measured precisely like reliability. (The exception is criterion validity, which can be reported as a correlation, $-1 < r < 1$)

The 3 Classic Types of Test Validity

Type of Validity	Description	How to Get It
Construct	The degree to which the measure or test measures the construct. Needs to show that it's similar to and different than other measures.	**Convergent validity:** Demonstrate that the measure is similar to other measures that are theoretically similar. **Example:** A new, easy to use pre-employment test that measures manual dexterity. **Divergent validity:** Demonstrate that the measure isn't measuring something else. **Example:** work performance vs cognitive ability vs conscientiousness vs manual dexterity.
Criterion	The degree to which the measure or test score is correlated to the desired outcome (e.g., productivity or performance).	**Concurrent Validity:** Calculate the correlation between present workers' test scores and their performance. **Predictive Validity:** Calculate the correlation between applicants' pre-employment test scores and their performance once they are employed.
Content	The degree to which all aspects of the construct are measured.	Ask an expert about the construct. Study the construct to discover all that's associated with it.

Note:
Establishing reliability and validity of a measure is difficult. If possible, use measures that already have established levels of reliability and validity; don't try to reinvent the wheel!

Example: If you want to measure the Big Five personality traits, consider using the Big Five Inventory (44 item version) (BFI; John & Sirvastava, 1999).

Chapter 6. Experiments

Experiments and Causation

How do we demonstrate causation?

Does a treatment/condition influence the outcome?

What is necessary to demonstrate causation?
1. -The cause (IV) and effect (DV) vary together (beyond chance)
2. -The cause (IV) precedes the effect (DV)
3. -There are no other plausible explanations on why they vary together.
   -Example: Ice cream consumption and crime

The essential nature of an experiment is **random assignment** of participants into groups.
   -Each group receives a different treatment (the IV).
   -If there is any difference between groups concerning the DV beyond what we would expect by chance, it is probably due to the different treatments (the IV).

Correlation does not imply causation.

If two variables A and B are correlated, there are 4 causal interpretations:
1. A causes B.
2. B causes A.
3. A third factor causes both A and B.
4. The correlation is spurious; it is due to chance.

Discussion

1. What are some examples of correlation where one variable causes another?

2. What are some examples of correlation where a third variable causes both variable A and variable B?

3. When might Christians be especially susceptible to assuming that correlation implies causation?

Stroop Experiment Instructions

On the following pages are materials necessary to perform a classroom experiment demonstrating the Stroop effect. You can read about the Stroop effect on Wikipedia or any Intro to Psych textbook.

Required Documents
These can be downloaded from https://osf.io/x7ruy/ . Some of them need to be printed in color.
>	Chapter 6 Stroop Colors.docx
>	Chapter 6 Stroop Colored Squares.docx
>	Chapter 6 Stroop Colored Words.docx
>	Chapter 6 Stroop Black Words.docx
>	Chapter 6 Stroop Effect Experiment Worksheet.docx

Directions
1. Randomly assign members of the class (up to 20 people) into two groups using the numbers on the file **Chapter 6 Stroop Numbers for Random Assignment.docx**.

2. The people who chose odd numbers (the odds) should sit on one side of the classroom and the people who chose even numbers (the evens) on the other. Place a table in the front and center of the classroom that everyone can see.

3. Distribute a copy of **Chapter 6 Stroop Effect Experiment Worksheet.docx** to everyone in the class.

4. Show each group a color copy of the sheet **Chapter 6 Stroop Colors.docx**. Depending on the printer used, some of the colors may be controversial. However, use the color names as defined on this sheet.

Part 1
5. In **Part 1** of the experiment, the odd group will be the participants first. They will be the control group. The evens need to select one person to be the experimenter who will run the odds through experiment, one person to be the timer, and one person to record the results on Excel (or SPSS) which should be broadcast onto a screen.

5. The experimenter calls participants from the odd group one at a time to come sit at the table in front of the classroom. He or she puts the participant at ease and explains that the participant will look at a sheet of 20 colored squares and state what color each square is.

6. After the participant understands what he or she will do, the experimenter turns over the sheet **Chapter 6 Stroop Colored Squares.docx**. The timer records how long it takes to identify the 20 colors. The score ($O_{2Control}$) is recorded on Excel so that the whole class can see the score. The experimenter may give the participant a piece of candy as a means of thanking him or her for participating.

7. After all of the participants have provided scores, calculate the average time $O_{2Control}$ and record it on the worksheet.

8. Now the evens will be the experimental group and the odds will provide the experimenter, the timer, and someone to record the scores for $O_{2Experimental}$. Complete the data collection for the experimental group as in steps 5.-7., except use the sheet **Chapter 6 Stroop Colored Words.docx**. The participants are to identify the color of each word (not what the word says).

9. When the means of $O_{2Control}$ and $O_{2Experimental}$ have been calculated, graph them on the x-y axes provided on the worksheet. Then calculate t, record the df, and note the p value. Discuss its meaning.

Part 2
10. In **Part 2** of the experiment, you will test whether the color of words interferes with the meaning of the words. This is similar to the Stroop effect, but it is more difficult to detect. Therefore, we will use a more powerful experimental design: a Pretest-Posttest Control Group.

11. The evens will be the participants first, as the control group. Using the same procedure as before, members of the odd group will first record $O_{1Control}$ using the sheet **Chapter 6 Stroop Black Words.docx**. Immediately after all of the evens have provided a score for $O_{1Control}$, they should repeat the procedure for $O_{2Control}$. There is no difference in the pretest and posttest conditions (there is no treatment). Calculate and record Mean $O_{1Control}$ and Mean $O_{2Control}$.

12. The odd group will be the experimental group. For $O_{1Experimental}$, use **Chapter 6 Stroop Black Words.docx** as with the control group and calculate. However, use **Chapter 6 Stroop Colored Words.docx** for the experimental condition. Record $O_{2Experimental}$.

13. Calculate and record Mean $O_{1Experimentall}$ and Mean $O_{2Experimental}$. Make a bar chart of all 4 means on the axes provided.

14. Calculate the difference between the pretest and the posttest scores for each participant. Calculate Mean Change$_{Control}$ and Mean Change$_{Experimental}$. Then calculate t for these two means, record the df, and note the p value. Discuss its meaning. (Note: This analysis can be performed using ANOVA; choose the analysis most appropriate given the stats ability of the class.)

Stroop Effect Experiment Worksheet

Part 1: Semantic Interference

Research Question: Does the meaning of words (the semantic meaning) interfere with identifying the color of something?

For example, is it harder to identify the color of something printed in black when it's in the form of a word "red" rather than a neutral shape (■)?

H_1: Semantic meanings interfere with color identification.

Experimental Design: Posttest-only control group

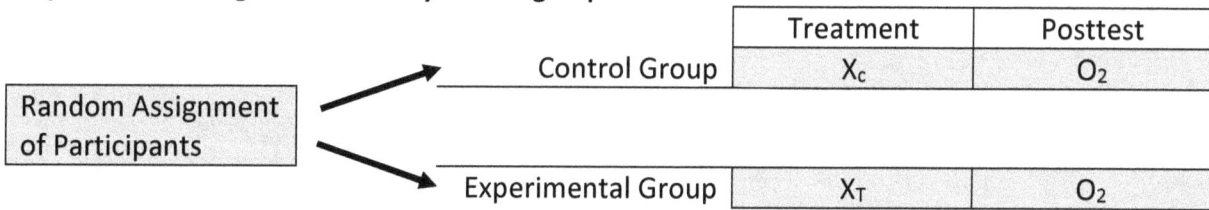

	Treatment	Posttest
Control Group	X_C	O_2
Experimental Group	X_T	O_2

X_C Treatment received by control group: Presentation of colored shapes
X_T Experimental treatment: Presentation of colored words
O_2 Posttest measurement of time needed to identify colors

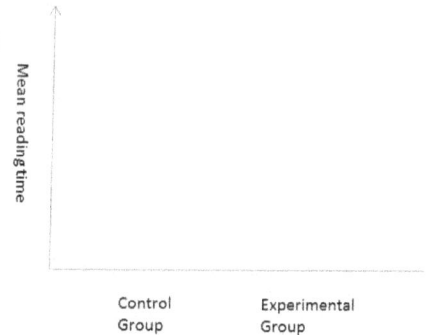

Results: Mean $O_{2control}$ =   Mean $O_{2Experimental}$ =

Is the difference significant (Independent two-sample t test)?

  t = df = (   ) p =

Part 2: Color Interference

Research Question: Does the color of words interfere with the semantic meaning of words?

H_2: Color interferes with the semantic meaning of words

Experimental Design: Pretest-Posttest Control Group

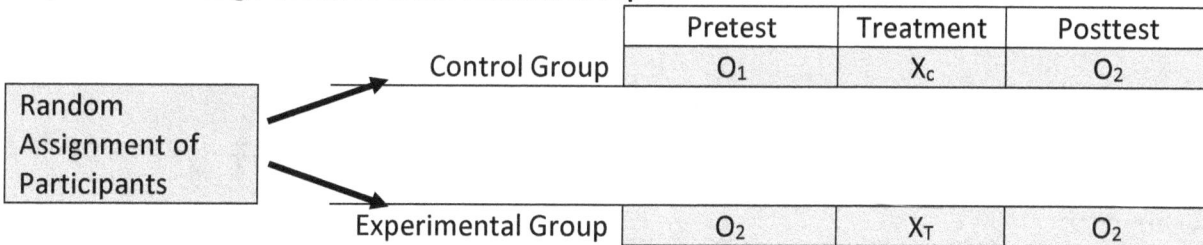

	Pretest	Treatment	Posttest
Control Group	O_1	X_C	O_2
Experimental Group	O_2	X_T	O_2

O_1 Pretest measurement of reading time for grey words
X_C Treatment received by control group: Gray words remain grey
X_T Experimental treatment: Gray words are replaced with color words
O_2 Posttest measurement of reading time

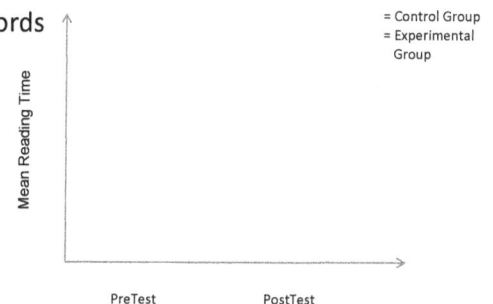

Results: Mean $O_{1Control}$ =   Mean $O_{2Control}$ =

           Mean $O_{1Experimental}$ =   Mean $O_{2Experimental}$ =

           Mean Change$_{Control}$ =   Mean Change$_{Experimental}$ =

Is the difference in change significant (Independent two-sample t test)?

  t = df = (   ) p =

Numbers for Random Assignment

This sheet can be downloaded from https://osf.io/x7ruy/

1

2

3

4

5

6

7

8

9

10

11

12

13

14

15

16

17

18

19

20

Print and cut the numbers into small tickets.
Place them in a hat or envelope and have participants select one in order to be assigned to one of two groups, **odd** or **even**.

Colored Squares for Stroop Experiment

This sheet needs to be downloaded from https://osf.io/x7ruy/ and printed in color.

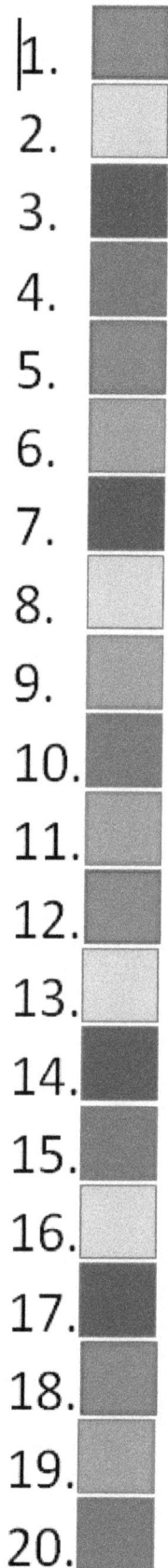

1.
2.
3.
4.
5.
6.
7.
8.
9.
10.
11.
12.
13.
14.
15.
16.
17.
18.
19.
20.

Stroop Black Words/Stroop Colored Words

Both the colored and black words can be downloaded at https://osf.io/x7ruy/

1. GREEN
2. BLUE
3. ORANGE
4. GREEN
5. PURPLE
6. ORANGE
7. BLUE
8. RED
9. BLUE
10. PURPLE
11. RED
12. BLUE
13. PURPLE
14. GREEN
15. ORANGE
16. GREEN
17. RED
18. ORANGE
19. PURPLE
20. RED

Chapter 7. Presenting Scientific Research

Basic Outline of a Research Report or Thesis

There are 4 main sections:

I.   Introduction to the problem.
    A.   Broad presentation of the need.
    B.   Review of the literature (what we already know about the problem and potential solutions).
    C.   Hypotheses.
II. Methods (What we did to get additional information).
III. Results (The information that we gained and its meaning).
IV. Discussion (How the results relate to the problem and to previous research).

Detailed Outline for a Research Report or Thesis

This is a typical outline for an APA journal article, a research report, or a thesis in psychology

I. Title Page
II. Abstract (Executive Summary)
III. Introduction
    A. Description of the problem and its importance
    B. Review of previous research (Lit review including definitions of important concepts)
    C. Purpose of the present research including theoretical framework used
    D. One or more hypotheses
IV. Method
    A. Introductory overview
    B. Participants
        1. Characteristics
        2. How the sample was obtained
    C. Measures (of each variable)
        1. Questions used
        2. How computed
    D. Procedures (used to collect data)
V. Results
    A. Descriptive statistics
        1. Mean and standard deviation of each variable
        2. Intercorrelation table
    B. Tests of the hypotheses
    C. Qualitative/Subjective results
VI. Discussion
    A. Summary of the study
    B. Meaning and Interpretation (Be convincing!)
        1. In light of past research
        2. In light of hypotheses
    C. Implications
        1. New information gained
        2. Recommendations
    D. Limitations
VII. References
VIII. Tables (include in body of text when not submitted for publication)
IX. Figures (include in body of text when not submitted for publication)
X. Appendices (optional)
    A. Survey or test (not optional for your study)
    B. Computer output
    C. Additional tables and charts
    D. Raw data

Evaluating Your Research Report or Thesis

Title Page
1. Is everything APA formatted, according to the most recent APA Manual?
2. Does the title clearly (and perhaps even cleverly) explain the study?

Abstract
1. Does your abstract clearly summarize the research problem, the hypothesis, the methods used, who the participants are, and the results?
2. Do you summarize the importance of the study?

Introduction
1. Is the problem statement clear?
2. Does the lit review relate clearly to the problem statement?
3. Does the lit review cite all the fundamental, classic research relevant to the topic?
4. Does the lit review cite the most recent studies that are directly relevant to the topic?
5. Does the lit review present a clear and coherent argument that leads to the hypotheses?
6. Are the hypotheses stated clearly and unambiguously?
7. Do the hypotheses follow logically from your problem statement and your lit review?

Method
1. Does the introductory overview present the big picture of your research approach?
2. Is each variable clearly introduced with a description of what it measures?
3. Are sample items for each measure included?
4. Is the sampling method described clearly in such a way that others could reproduce it?
5. Are the characteristics of the participants clearly explained so that the reader has a good sense of who is included in the sample?
6. Are the procedures used to obtain the data clear enough for other researchers to repeat this study?

Results
1. Are all the results of the statistical analyses presented in APA format?
2. Is the significance of each statistical analysis explained?
3. Are the results linked to each of the hypotheses?
4. Are graphs and tables presented in APA format?

Discussion
1. Are the results summarized without using any numbers?
2. Are the results linked to past research?
3. Are the results linked to the research problem and the hypotheses?
4. Are the practical implications discussed?
5. Are the limitations to the study presented, as well as additional research that should be carried out?

Appendices
1. Is a complete version of your survey instrument attached?
2. Is all additional information that would potentially interest a reader attached?

Research Proposals vs. Research Reports

Significant research is usually preceded by a research proposal which explains the research you plan to do: both the **why** and the **how**.

Section	Proposal	Final Report/Thesis
Introduction	A complete description of the problem statement, an extensive lit review, and the hypotheses.	Same as in the proposal. Only the relevant aspects of the lit review should be included.
Method	A complete description of what you are planning to do. Written in the **future** tense.	A complete description of what you actually did. Written in the **past** tense.
Results	A brief summary of the analyses you will carry out. Written in the **future** tense.	A complete presentation of all the analyses and results. Written in the **past** tense.
Discussion	A brief summary of what you think the significance of this study will be. Written in the **future** tense.	Discussion of actual results and their significance. Written in the **past** tense.
Appendices	The survey Instrument. IRB documents, if necessary.	The survey instrument. All other relevant information.

Resisting the Temptation of Unethical Practices

All researchers face temptations varying from plagiarism to falsifying data. Read the following article *Fraud Scandal Fuels Debate Over Practices of Social Psychology* which originally appeared in the *Chronicle of Higher Education*:

http://www.chronicle.com/article/As-Dutch-Research-Scandal/129746/

Discussion

1. This article argues that psychology "has become addicted to surprising, counterintuitive findings that catch the news media's eye, and that trend is warping the field."
Do you think this is true? Would it be truer now than in the past?

2. In paragraph 4, several of Stapel's papers are mentioned. Why would these be such hot subjects? How should Christians respond to hot topics?

3. In what ways could you cheat in your current research?

4. What Christian principles are especially relevant to honesty in research?

5. What specific steps can you take to ensure that you don't succumb to the temptations associated with cheating?

Oral Presentations

When presenting research before others, it's important to follow all of the guidelines provided by the organizer.

1. Time of Presentation: Academic conferences often have a referee who will provide 2 minute warnings and will then cut off the speaker if he or she does not finish on time.

2. Practice: Your presentation will be smoother if you practice it several times before the presentation.

3. Proofreading: Errors and typos are unnecessary distractions that make the presenter seem less competent.

4. PowerPoint Presentations:
    a. Use a consistent slide theme that is not distracting.
    b. Tables and figures are usually helpful, unless they are too complicated or inappropriate for the audience (e.g., "cutesy" pictures, crude humor, etc.).
    c. Use large fonts (at least 24 point) on your slides. Try not to put more than 7 lines of text on a slide. Do not put entire paragraphs.
    d. Don't read long passages off your slides. Speak with passion and enthusiasm.

5. Keep the focus on you, the presenter:
    a. Don't turn off the lights so that it's difficult to see your facial expressions.
    b. Communicate the value and meaning of your information through your vocal inflections, facial expressions, and gestures.
    c. Track what your audience is feeling; adjust the presentation appropriately.

Discussion

1. What do you think is most likely to go wrong with your group's oral presentation of your research project?

2. What can you do to reduce the likelihood of these problems from occurring?

(Special thanks to Wendi Dykes of Azusa Pacific University for many of these ideas.)

Chapter 8. Qualitative Research Methods

Qualitative Research vs. Quantitative Research

(See *Qualitative Inquiry & Research Design* by John Creswell)

I. Definition of Qualitative Research
    A. "Collecting, analyzing, and interpreting subjective (non-quantitative) data to persuasively answer a research question concerning an important issue."
    B. Key parts
        1. Collecting data
        2. Analyzing data
        3. Interpreting data
        4. Subjective (non-quantitative) data
        5. To answer a research question
        6. Persuasively
        7. Concerning an important issue

II. Typical Characteristics of Qualitative Research (vs. quantitative research)
    A. A natural setting (vs. laboratory setting)
    B. The researcher develops a unique instrument/interview questions/source of information to collect data (vs. psychometrically validated measures of specific concepts).
    C. Multiple methods (interviews, focus groups, observations, surveys) (vs. a survey or data from an experiment)
    D. Focus on participants' meaning (vs. psychometrically validated measures of specific concepts).
    E. Emergent design (vs. predetermination of a design to reduce bias).
    F. Reflexivity (vs. attempts at objectivity)

III. The Purpose of Qualitative Research (vs. quantitative research)
    A. Exploring a problem or developing a theory, rather than testing a hypothesis: to show what's important

Discussion

1. What are the advantages and disadvantages of qualitative (vs. quantitative research)?

2. Looking at the definition, what aspects of qualitative research seem the most difficult?

3. What natural settings do you have access to?

Types of Qualitative Research

Characteristic	Narrative	Phenomenology	Grounded Theory	Ethnography	Case Study
Goal	Explore the life of an individual	Understand the essence of an experience	Develop theory grounded in data	Describe / interpret a culture-sharing group	Develop in-depth description & analysis of a case
Result	Stories of the person's **experiences**	**Description of** how people have lived a **phenomenon**	A **theory** grounded in the views of the participants	**Description** and interpretation of a group's **culture**	**Application** based on an example
Unit of Analysis	An individual (n = 1)	A Phenomenon (among individuals with the same experience, n = about 30)	A process, action, or interaction (n = about 30)	A culture-sharing group (multiple individuals will be observed)	An event, program, or activity (typically more than one individual is involved)

Key textbook (easy to read, great examples):
Creswell, J. W. (2013). *Qualitative inquiry and research design: Choosing among five approaches*. Sage.

Types of Data Used in Qualitative Research

I. Interviews
   A. Questions need to be developed
      1. Can be fixed to get consistent data sets.
      2. Can vary as the interview or study evolves.
   B. Should be recorded
   C. Should be transcribed
II. Focus Groups
   A. Essentially group interviews
   B. Emotionally and cognitively different dynamics than individual interviews
III. Surveys
   A. Most common way of gaining quantitative data in organizations.
   B. Not great at getting qualitative data, but potentially useful if participants are motivated.
IV. Observations
   A. Recorded via note taking, photos, or videos.
V. Documents
   A. Official documents
   B. Historical documents
   C. Advertising
   D. Any type of relevant document
VI. Audiovisual materials
   A. Photos, recordings, videos, etc. already in existence.

Discussion

1. If you were to redo your study as a qualitative study, what type of data sources do you think would be most valuable?

2. Which data sources do you think are the most subjective? Which do you think are the most objective?

3. In sermons, is there more quantitative or qualitative data presented? Why would this be? Do you think it should be this way?

Narrative Studies and Case Studies

Narrative and Case Studies: Examining one person, organization, or event as in-depth as possible.

I. **Advantages**
    A. Focusing on only one person, organization, or event allows for a more in depth study
    B. It incorporates a variety of methods
        1. Literature review
        2. Personal observation
        3. Interviews
        4. Archival research
        5. Anything else that the researcher can think of.
    C. Results are interesting.
    D. It identifies further problems to study.
II. **Disadvantages**
    A. It is difficult to do well, easy to do poorly.
        1. It can be very time consuming and intrusive.
        2. It requires insight and excellent writing skills (rich descriptions).
    B. It is very subjective
        1. Observations are biased by the observer's biases.
        2. Little reliability; another researcher might come to a very different conclusion.
    C. There is no way to detect causation
        1. No experimental control is possible.
    D. There is little external validity
        1. $N = 1$
        2. We can't generalize conclusions without a larger sample.
III. **Conclusion:** Narrative and case studies are great for figuring out what hypotheses should be tested.

Exercise

1. Choose one person in your group who excels in some way. For example, someone who is a champion swimmer or who knows how to make very large bubble gum bubbles.

2. Do a mini-case narrative study of this person to explain how the person came to excel. Propose reasons why this person excels in the given domain. Do not limit yourself to reasons that the person in question proposes.

3. Write down at least one testable hypothesis concerning the skill or ability in question that flows out of your narrative study.

Searching for Physical Artifacts

With your research team, look around your school for meaningful artifacts (physical artifacts or elements of the environment) that show evidence that your school is or is not accomplishing its mission.

For example, the mission of Azusa Pacific University is:

Azusa Pacific University is an evangelical Christian community of disciples and scholars who seek to advance the work of God in the world through academic excellence in liberal arts and professional programs of higher education that encourage students to develop a Christian perspective of truth and life.

These artifacts may either:

    a. Provide evidence concerning the degree to which your school is accomplishing its mission.

    b. Indicate values or practices that should be affirmed or modified in light of the mission statement.

Exercise

1. Choose several artifacts that are interesting and relevant to the school's mission. Discuss them with your team until you agree on their meaning.

2. Take a picture of each artifact.

3. Come back to class and present your artifacts in a slide show. Discuss their meaning in light of the school's mission.

4. Practice "reflexivity": Discuss what your own biases are and how this influenced your choice of artifacts and your interpretation of them.

Focus Groups

A focus group:

- A group gathered and moderated by a researcher for generating new information relevant to the problem or issue.
- Can be very productive, but facilitators must:
  - Keep group focused on task of gaining useful information.
  - Remain neutral.

Purpose	Example
Collect information (**To understand behavior, opinions, and perceptions**)	Ask a group of engineers "How useful do you think it would to be have a once a week meeting simply to update each other on the projects you are working on?"
Determine what people think in a group (which might be different from what they would say as individuals)	Ask a group of younger employees "We'd like to know how you feel as a group about this new benefits package that we're considering proposing to new employees?"
Generate insight into perceived causes (of behavior, opinions, and perceptions)	Ask a group of manufacturing employees "This last year absenteeism has been higher than in the past. What do you think might be some of the causes of this?"
Determine how a group functions or how they make decisions (to understand their motivation and processes)	Ask a group of firemen "How did you go about deciding how new information would be shared with a new unit when they arrive at an active fire?"

Focus Group Exercise

Problem: We want to improve student life for students in our program.

Divide into two groups. Students and Administrators. While the Administrators work on developing questions to understand how student life can be improved, the students can talk about what they have experienced as students in their program.

Once the Administrators are ready, they should arrange the room appropriately and get whatever information they can from the focus group of Students.

Coding Interviews

An excellent exercise in coding interviews for qualitative research can be found at:
http://humbox.ac.uk/4052/

It is part of the HumBox project of the Open Educational Resources initiative in the UK.

An additional transcribed interview for this exercise can be downloaded at:
https://www.academia.edu/2720752/Qualitative_Research_Methods_Coding_Exercise

Grounded Theory and Interviewing

Purpose Statement

The purpose of this grounded theory study is to develop a theory of the influence of early work experiences on one's attitude towards organizations. The study will focus on students in your school. At this stage in the research, early work experiences will be defined as the first several paid employment experiences held by the participants before they graduated from college.

Research Question

How do early work experiences influence a person's attitude toward organizations later in life?

Subquestions:

1. What are the characteristics of people's early work experiences?
2. What are people's current attitudes towards organizations?
3. What early work experiences influenced people's current attitudes?

Suggested Interview Questions

1. Tell me about your **first work experience**.
    a. Introductory questions
        i. Where was it?
        ii. What organization?
        iii. What were your responsibilities?
        iv. How did you get this job?
        v. How much did you make?
        vi. How long did you stay?
        vii. Why did you leave?
    b. Describe the personality of your boss.
        i. Did you enjoy working for him or her?
            1. Explain an interaction (or other event) with your boss that you enjoyed or did not enjoy.
        ii. How fair did your boss treat you?
            1. Can you give an example of how he treated you fairly or unfairly?
            2. How did he treat the other employees?
        iii. How would you describe your relationship with your boss?
            1. Tell me a story that illustrates this.
    c. Who were your coworkers?
        i. Who did you like the most?
            1. Why?
            2. Describe something that this person did.
        ii. Who did you like the least?
            1. Why?
            2. Describe something that this person did.
        iii. Who else do you remember?
        iv. What other memories do you have of your coworkers in terms of:
            1. Relationships?
            2. How they treated you?
            3. How you interacted with them?

  d. What were your responsibilities?
    i. How competent did you feel you were for each one?
    ii. Which ones did you enjoy the most?
    iii. Which ones did you enjoy the least?
    iv. What did you learn about the task?
      1. How useful has this been for you?
    v. What training was provided?
      1. How useful was this?
  e. Overall, how useful was this work experience for your growth?
    i. Your personal growth?
    ii. Your professional growth?
  f. To what degree would you like other young people to have this sort of experience?
    i. What would be most useful?
    ii. What would be least useful?

2. Now I want to explore **your attitude toward organizations**. You're a student in a field that has a lot to do with organizations, so I imagine you've thought a lot about organizations.
  a. What type of organizations do you appreciate the most?
    i. Can you cite some examples?
    ii. What experience do you have with these types of organizations?
  b. With what type of organizations have you had the most experience?
  c. In general, how well do you think organizations do what they are supposed to do?
    i. Please provide me an example illustrating this.
  d. In general, how good to people do you think organizations are?
    i. Please provide me an example illustrating this.
  e. What do you think is the biggest need in organizations?
    i. Can you describe a situation where you've seen this need?
  f. What do you think is the greatest strength of organizations?
    i. Can you tell me about something that you've seen that illustrates this?

3. How do you think your first job influenced your present attitude toward organizations?

4. *(If time)* Do the section 1 and 3 questions for **the person's 2nd and 3rd job**. Add, modify, or skip any questions as appropriate.

Exercise

The goal of this exercise is to give you practice interviewing, recording, transcribing, coding, and interpreting qualitative data for a grounded study. It will take several days to complete.

1. Learn how to record an interview on your phone or some other digital recorder. Become an expert in the process, including placement of the microphone so that the interview can be recorded without technical difficulties.

3. For 30-45 minutes interview a person in the class using the above interview.

4. Transcribe the interview.

5. Code the interview for relevant themes.

6. Make copies of the coded transcripts for everyone on your team. Together, try to synthesize your findings and create a grounded theory concerning one's early work experiences and their attitudes towards organizations.

Index

www.ingramcontent.com/pod-product-compliance
Lightning Source LLC
Chambersburg PA
CBHW081647270326
41933CB00018B/3381